Showdown

Animal Groups

Heather E. Schwartz

Publishing Credits

Rachelle Cracchiolo, M.S.Ed., *Publisher*
Conni Medina, M.A.Ed., *Managing Editor*
Nika Fabienke, Ed.D., *Series Developer*
June Kikuchi, *Content Director*
John Leach, *Assistant Editor*
Kevin Pham, *Graphic Designer*

Image Credits: pp.8, 9, 16 bottom: Martin Strmiska/Alamy; pp.10, 11, 16 middle: Ivan Vdovin/Alamy; all other images from iStock and/or Shutterstock.

Library of Congress Cataloging-in-Publication Data

Names: Schwartz, Heather E., author.
Title: Showdown : animal groups/ Heather E. Schwartz.
Description: Huntington Beach, CA : Teacher Created Materials, [2018] | Audience: K to Grade 3.
Identifiers: LCCN 2017026877 (print) | LCCN 2017030011 (ebook) | ISBN 9781425853280 (eBook) | ISBN 9781425849542 (pbk.)
Subjects: LCSH: Animals--Habitations--Juvenile literature.
Classification: LCC QL756 (ebook) | LCC QL756 .S39 2018 (print) | DDC 591.56/4--dc23
LC record available at https://lccn.loc.gov/2017026877

Teacher Created Materials
5301 Oceanus Drive
Huntington Beach, CA 92649-1030
http://www.tcmpub.com

ISBN 978-1-4258-4954-2

Many animals share their homes. They live in groups.

Toucans live in the forest.
They live in groups called flocks.

Wolves live in the woodlands.
They live in groups called **packs**.

Dolphins live in the ocean.
They live in groups called pods.

Zebras live in the grassland.
They live in groups called **herds**.

Penguins live on the ice.
They live in groups called **colonies**.

Many animals share their homes. They live in groups.

That's just like you and me!

Glossary

colonies

flocks

herds

packs

pod